ELEMENTS OF THE PERIODIC TABLE

WHAT IS SILVER?

047

Ag

SILVER

107.87

KATHLEEN A. KLATTE

ROSEN PUBLISHING

Published in 2026 by The Rosen Publishing Group, Inc.
2544 Clinton Street, Buffalo, NY 14224

Portions of this work were originally authored by Brian Belval and Jun Lim and published as *Silver*. All new material this edition authored by Kathleen A. Klatte.

Designer: Rachel Rising
Editor: Kathleen A. Klatte

Library of Congress Cataloging-in-Publication Data
Names: Klatte, Kathleen A., author.
Title: What is silver? / Kathleen A. Klatte.
Description: Buffalo, New York : Rosen Publishing, [2026] | Series: Elements of the periodic table | Includes bibliographical references and index.
Identifiers: LCCN 2024062082 | ISBN 9781499478938 (library binding) | ISBN 9781499478921 (paperback) | ISBN 9781499478945 (ebook)
Subjects: LCSH: Silver--Juvenile literature. | Chemical elements--Juvenile literature. | Precious metals--Juvenile literature.
Classification: LCC QD181.A3 K53 2026 | DDC 546.654--dc23/eng/20250212
LC record available at https://lccn.loc.gov/2024062082

Manufactured in the United States of America

CPSIA Compliance Information: Batch #CSRYA26. For further information, contact Rosen Publishing at 1-800-237-9932.

CONTENTS

INTRODUCTION

Silver (Ag) is a soft, silver-gray metal that's been prized by humans for millennia. Silver jewelry dating back as far as 4400 BCE has been found in ancient Egyptian tombs. The Romans began producing silver coins in the third century BCE. For centuries, silver has been used to produce coins, jewelry, and beautiful decorative objects such as candelabras and tableware.

In the 16th century, glass makers in Venice, Italy, found a new way to make mirrors by coating the back of glass with silver. Although mirrors weren't a new invention, using silvered glass to produce them was. These new "looking glasses" were popular among royalty and the very wealthy.

If you visit a museum, you might see antique silver objects up close. You might also see paintings depicting silver objects in use in wealthy homes. For centuries, the only way to capture images was to have a picture of a subject painted by an artist. This was a time consuming and expensive process. Only the very wealthy could afford to have their portraits painted.

That changed in the 1830s, when two Frenchmen, Joseph-Nicéphore Niépce (1765–1833) and Louis Daguerre (1789–1851) invented ways of taking the first photographs with the help of silver. Their invention involved some interesting chemistry.

Silver iodide (silver combined with iodine) darkens when exposed to light. To create photographs, they coated a piece of metal with a thin layer of silver iodide. The metal with its coat, known as a plate, was placed inside a type of camera made from a box with a small hole in it. When the hole was uncovered, it would allow light to enter the camera and darken the silver iodide on the plate. The result was a black-and-white photograph of whatever was in front of the camera.

Although photography lessened the time required to sit for a portrait, it still took time for the image to develop. Today's photographs happen using a different, more modern process that is faster.

Early photography was still expensive and very complex. Cameras were large, the photographer needed to carry a supply of toxic chemicals, and the photographic plates were very difficult to handle. About 50 years later, American inventor George Eastman (1854–1932) invented roll film. This type of film was also coated with silver iodide, but it was far easier to use. Eastman's innovations helped make photography accessible to the public.

Silver ink is used to manufacture radio-frequency identification (RFID) tags. You've probably seen these tags on merchandise in stores. They're used to deter theft.

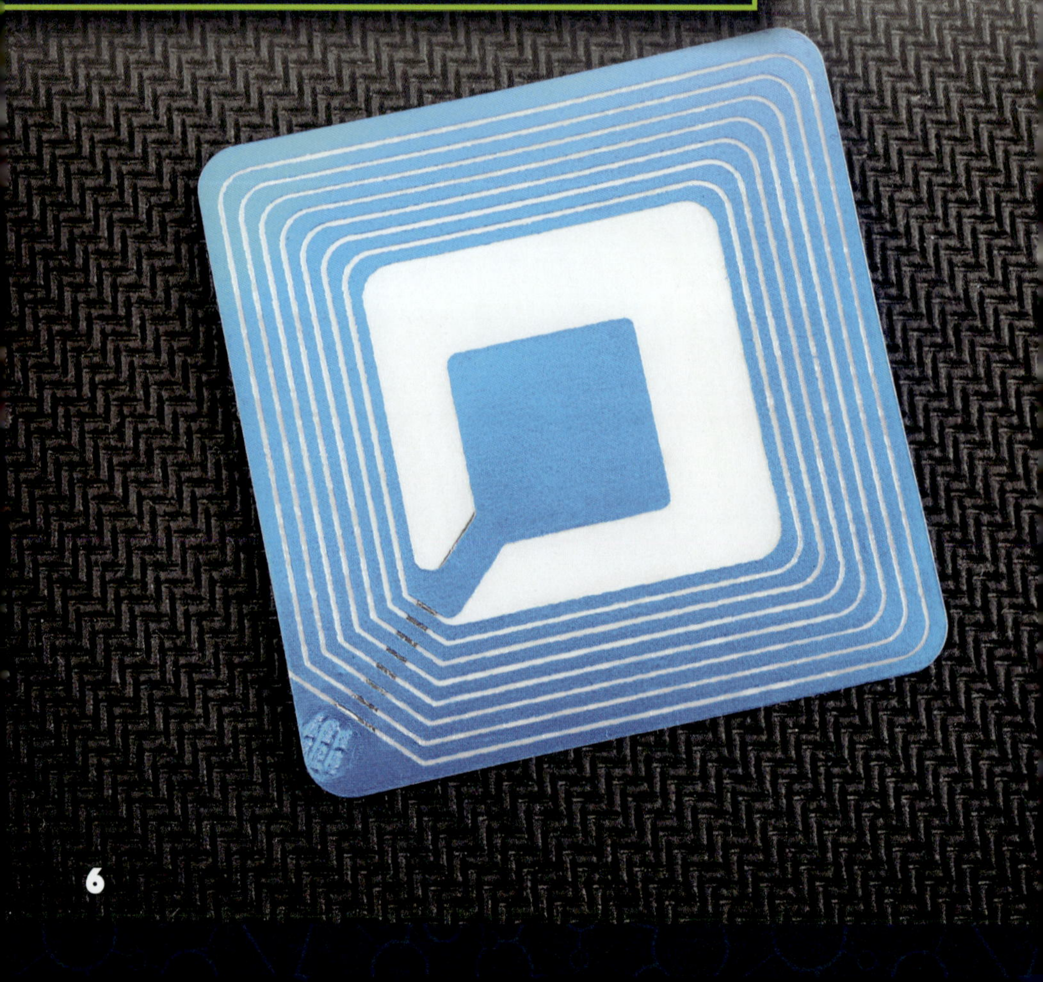

Although gold has a much higher dollar value today, solid silver flatware and serving pieces can fetch good prices from a jeweler or antique dealer.

Today, people still wear silver jewelry. However, silver coins are mostly collector's items rather than everyday currency. Most photography today is digital, and modern decorating trends have veered away from silver objects that need to be polished to look their best. Silver is still present in our everyday lives, though. Silver is used in batteries, 3-D printing, solar panels, and electronics. Silver is prized in the medical community for its antimicrobial properties. Even if you can't spot it, there's probably silver in things in your home that you use every day, so it's good to know more about this element.

CHAPTER ONE

A SILVER PLACE SETTING

When you think of silver, you might imagine jewelry, or perhaps an antique silver teapot that your grandparents display in a fancy cabinet. However, silver isn't just a familiar metal. It's one of over 100 chemical elements that make up everything in the universe. An element is a substance made up of all the same kind of atoms. A water (H_2O) molecule, which is a compound, can be broken into hydrogen (H) and oxygen (O) atoms. A silver molecule is only composed of silver atoms. People have valued silver since the earliest civilizations. It's still precious today for different reasons. Let's find out why!

Elements on the periodic table are listed in order of their atomic number. Silver's atomic number is 47 because that's the number of protons in its nucleus.

WHAT ARE ATOMS MADE OF?

Silver, like all matter, is made of tiny particles called atoms. Atoms themselves are made of even smaller particles: electrons, protons, and neutrons. At the center of every atom is the nucleus. The nucleus consists of protons and neutrons packed tightly together. Protons have a positive charge, and neutrons have no charge. The nucleus serves as the solid core of the atom. A silver atom always has 47 protons in its nucleus. This is what makes an element different from all the other elements: Each has a different number of protons. For example, a hydrogen atom has one proton in its nucleus, while a gold atom has 79.

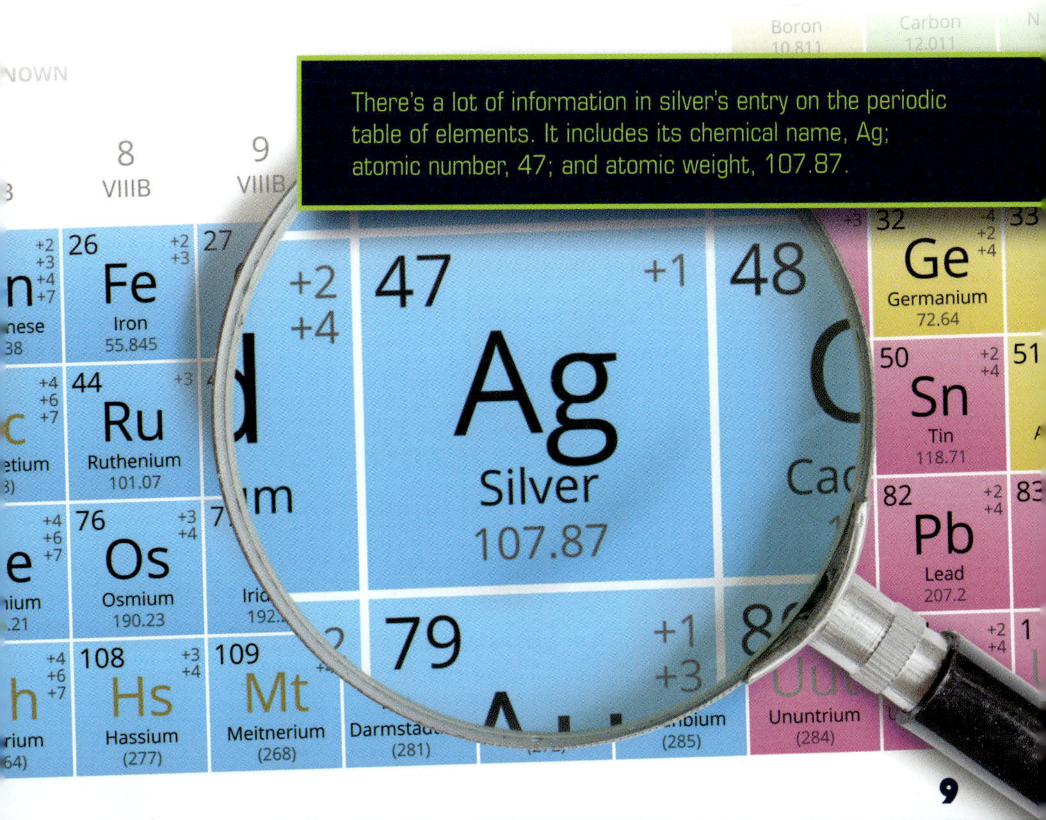

There's a lot of information in silver's entry on the periodic table of elements. It includes its chemical name, Ag; atomic number, 47; and atomic weight, 107.87.

Electrons are negatively charged and are attracted to the positively charged protons. The attraction is similar to the way opposite poles of a magnet attract each other. Electrons are extremely small and light—they weigh about 2,000 times less than a proton or neutron. The electrons dart around outside the nucleus in paths known as shells. Each atom can have multiple shells. The silver atom has five shells of electrons. Two electrons are found in the first shell, eight in the second, 18 in the third shell, 18 in the fourth shell, and one in the fifth. In all the elements, the electrons in the outer shell are known as the valence electrons. These electrons are of special interest to chemists because they allow atoms to interact with other atoms. This process is known as chemical bonding.

In each atom, the number of electrons is always equal to the number of protons. This is necessary to balance out the positive and negative charges. If an atom gains or loses an electron, it is known as an ion. An atom that has lost one or more electrons has a positive charge and is called a cation. An atom that has gained one or more electrons has a negative charge and is called an anion.

NOT QUITE THE SAME

Atoms always have an equal number of protons and electrons (47 in a silver atom), but the number of neutrons in an atom of an element can vary slightly. Often, there are more neutrons than either protons or electrons.

Atoms with different numbers of neutrons are known as isotopes. Silver has two isotopes: an atom with 60 neutrons and an atom with 62 neutrons.

Silver isn't the only element that has isotopes. Take carbon, for example. Carbon has a famous isotope known as carbon-14. This isotope of carbon has eight neutrons and is radioactive—it breaks down over a period of thousands of years. By comparing the amount of carbon-14 in a sample to the amount of carbon-12 isotope (carbon atoms with six neutrons), scientists can figure out the age of the sample. This method is known as carbon dating.

Silver, along with copper and gold, is part of group IB of the periodic table. All three metals are fairly inert and resist corrosion. They also conduct electricity well, so they're often used for wiring and electronics.

A PLACE AT THE TABLE

The periodic table is designed to help people better understand the elements. In the table, the elements are arranged into rows and columns. The rows are known as periods, and the columns are known as groups. Each period and each group have a number. The periods are numbered one to seven from top to bottom. The groups are numbered in one of two ways, depending on the version of the periodic table you are using. (Some tables number the groups using both methods.) In the first system, the groups are numbered 1 to 18, from left to right. The second numbering system uses Roman numerals (I through VIII) followed by a letter (A or B), except for the noble gases at the far right, which are in group O. In this numbering system, the elements in the groups marked with an A are known as the representative elements. In the representative elements, the Roman numeral in the group equals the number of electrons in the outer shell of an element in that group. For example, elements in group VA all have five electrons in their outer shell. Elements marked with a B are known as the transition metals. Unlike the representative elements, the number before B in each of these groups does not always equal the number of outer-shell electrons. Like the transition metals, the noble gas elements also have varying numbers of electrons in their outer shells.

Below the main table is a smaller table consisting of two rows and fourteen columns. This block is part of the main table but has been moved so the table isn't too wide to fit across two pages. The first row of the smaller table should go in between lanthanum and hafnium on the main table. These elements are known as the lanthanides. The second row of the smaller table goes between actinium and rutherfordium. These elements are known as the actinides.

Elements with similar characteristics are placed together on the periodic table. Silver is a metal, like nickel, copper, gold, platinum, and cadmium. Metals conduct heat and electricity well. They can also be worked into different shapes.

Each element occupies one square of the table. In group 11 (IB), row 5, you will find silver. The letter or letters that represent an element are its chemical symbol. Silver's chemical symbol is Ag, which is short for *argentum*, silver's name in Latin. A few of the other elements also have Latin abbreviations, such as gold (Au for *aurum*) and iron (Fe for *ferrum*).

In each square of the periodic table are two numbers. The smaller number on the top left is the atomic number. It is equal to the number of protons in an atom of the element. As you can see, the atomic number increases from left to right and as you go down the table. The larger number in either the top right or at the bottom of each square is the atomic weight. The atomic weight is the sum of the number of protons and the average number of neutrons in an atom of the element.

The atomic weight and atomic number can be used to estimate the number of neutrons in an atom of a specific element. This calculation isn't exact, but it gives you a good idea:

ATOMIC WEIGHT - ATOMIC NUMBER = AVERAGE NUMBER OF NEUTRONS
IN THE CASE OF SILVER: 107.868 - 47 = 60.868 NEUTRONS

According to this formula, an atom of silver has 60.868 neutrons. However, an atom can only consist of whole neutrons, not fractions. Therefore, the average number of neutrons in a silver atom is 61. In reality, as noted earlier, an atom of silver has either 60 or 62 neutrons. This formula can only approximate the number of neutrons.

SILVER SNAPSHOT

CHEMICAL SYMBOL: Ag

CLASSIFICATION: Transition metal

PROPERTIES: Conducts electricity, malleable, ductile

DISCOVERED BY: Has been known since prehistoric times

ATOMIC NUMBER: 47

ATOMIC WEIGHT: 107.87 atomic mass units (amu)

PROTONS: 47 **ELECTRONS:** 47 **NEUTRONS:** 60 or 62

DENSITY AT 68°F (20°C): 10.5 grams/cubic centimeter

MELTING POINT: 1,763°F (962°C)

BOILING POINT: 4,014°F (2,212°C)

COMMONLY FOUND: Earth's crust in silver ores such as argentite

Silver's chemical symbol shows the importance of Latin in early science.

MENDELEEV'S TABLE

The first version of the periodic table of elements was designed by a Russian chemist named Dmitri Mendeleev (1834–1907) in 1869. Mendeleev wanted to make it easier for his students to recognize the similarities between different elements. His original table listed elements in increasing order of atomic weight. When he organized them like that, he realized that there were gaps in the table. Most scientists of the time agreed that there were more elements to be discovered. Mendeleev theorized that they would fill those gaps. His theory was proved correct in his lifetime, when additional elements were identified that fit into gaps in the table.

Today there are 118 officially recognized elements in the periodic table. Today's table is organized by atomic number. The table is updated and published by the International Union of Pure and Applied Chemistry (IUPAC). Scientists today believe that all the naturally occurring elements have been identified. The elements still being discovered are the result of lab experiments utilizing radiation. IUPAC supervises the process of identifying new elements—such as recording multiple experiments that yield the same results and naming new elements in the same style as what's already listed on the table. This ensures that scientists around the world know that they're all talking about the same substance.

This reproduction of the original periodic table is part of a monument to Mendeleev in St. Petersburg, Russia.

CHAPTER TWO

WHAT MAKES SILVER UNIQUE?

All elements have properties that make them unique. These properties can be chemical or physical. Chemical properties include the number of protons, electrons, and neutrons. They also include things like how easily an element reacts or combines with other elements. Chemical reactions result in a different substance being formed. For example, when two oxygen (O) atoms and one carbon (C) atom bond, they form carbon dioxide (CO_2), a totally different substance.

An element's physical properties are things that can be observed without changing the element to a different substance. This includes characteristics such as color, hardness, phase at room temperature, and melting point. Melted silver is still silver. Melting it doesn't turn it into a different substance.

IT'S A PHASE!

Matter exists in three phases, or states: solid, liquid, and gas. You are already familiar with these three phases. Think about water. Water in its liquid phase is a refreshing drink. Water in its solid phase is ice, and water in its gas phase is steam. These are the three phases of the substance with the chemical formula H_2O (two parts hydrogen to one part oxygen).

A simple way to describe a substance is by its phase at room temperature, which is approximately 68°F (20°C). Water's phase at room temperature is liquid. Silver, like all metals except mercury, is a solid at room temperature.

Artists have been creating intricate silver objects for centuries. These objects are treasured for their material worth and their beauty.

Silver has a distinct color, which is shown here.

WHAT COLOR IS SILVER?

It might seem obvious, but one notable property of silver is its color. Silver has a shiny gray-white appearance. You might even say that silver has a silver color! This contrasts with the color of some other metals, such as reddish copper and yellowish gold. When polished, silver reflects light better than any other metal. Because of this property, silver is used to make high-quality mirrors.

WHAT CAN YOU MAKE FROM SILVER?

Malleability is the ability of a material to be pounded into sheets. Silver is one of the most malleable of all metals. A piece of silver can be beaten into a sheet that's 0.0025 inches (0.00635 cm) thick. These thin sheets of silver are known as leaves. Its malleability is also the reason that silver can be shaped into intricate pieces of jewelry and tools.

Ductility is the ability of a material to be stretched into wires. Silver is impressively ductile. Metals that are ductile are often used to make electrical wire. Copper is most common for this purpose, mainly because it is inexpensive.

Silver conducts heat and electricity well. Today, it's valued for its use in electronic devices like computers and cell phones.

HOW HARD IS SILVER?

Silver is a relatively soft metal. Because it is so soft, silver is often combined with other metals to harden it. These mixtures of metals are known as alloys. A well-known silver alloy is sterling silver, which is 92.5 percent silver and 7.5 percent copper. The addition of copper makes the metal less likely to bend or become misshapen. This is especially important in the manufacture of products like silverware. A pure silver knife or fork would be too soft to be used as an eating utensil.

Scientists measure hardness using the Mohs' scale. The scale ranges from one to 10, with one representing the softest substance (talc) and 10 representing the hardest (diamond). Silver has a value of 2.5 on the Mohs' scale, which makes it slightly harder than gold but softer than most of the other metals.

HOW WELL DOES IT CONDUCT HEAT AND ELECTRICITY?

Electricity is the flow of electrons through a substance. Silver conducts electricity because its sole outer-shell electron can easily move from atom to atom. When voltage is applied, the electrons are set in motion, creating an electrical current. Silver is a better conductor of electricity than any other element. This property makes it very valuable in the electronics and electrical industries. Later on, we will look at exactly how silver is used in these industries.

In nearly the same way that silver conducts electricity, it will also conduct heat. If you were to stir a cup of hot cocoa with a pure silver spoon, you would feel the spoon handle become hot very quickly, much more quickly than with a steel or plastic spoon. Like its ability to conduct electricity, silver is better than all the other elements at conducting heat.

GETTING HOT IN HERE!

The melting point is the temperature at which a substance changes from solid to liquid. Silver's melting point is 1,763°F (962°C). That is much hotter than your oven at home can get. This property also makes silver a valuable element. Silver can be used at high temperatures and still retain its strength and structural integrity. Boiling point is the temperature at which a substance changes from liquid to gas. Silver has an impressive boiling point of 3,924°F (2,162°C).

The kind of furnace used to melt metals runs hundreds of degrees hotter than the furnace that heats your home.

METALS VS. NONMETALS

METALS	NONMETALS
INCLUDE SILVER, GOLD, IRON, ETC.	INCLUDE CARBON, SULFUR, CHLORINE, ETC.
CAN BE HAMMERED INTO THIN SHEETS	BREAK APART IF HAMMERED
CONDUCT HEAT AND ELECTRICITY WELL	POOR CONDUCTORS OF HEAT AND ELECTRICITY
STRONG—CAN SUPPORT GREAT WEIGHT	WEAK—CAN'T SUPPORT MUCH WEIGHT
HARD—SOLIDS CAN'T BE CUT WITH A KNIFE	SOFT—SOLIDS CAN BE CUT WITH A KNIFE
MAKE A RINGING SOUND WHEN STRUCK	NO RINGING SOUND WHEN STRUCK

HOW DENSE IS SILVER?

Density is the measure of mass per volume. It is an immutable property of an element. It's similar to weight, but weight is the force of gravity on a substance, which can change. If you took silver on a spaceship with less gravity, the silver would weigh less. It would float, like astronauts do. The density would remain the same.

Density is usually measured in grams per cubic centimeter (g/cm^3). Something that is dense is compact—a lot of mass is packed into a small space. Silver has a density of 10.5 g/cm^3. Aluminum (Al) is another common metal. It has a density of 2.7 g/cm^3. The least dense metal of all is lithium (Li) at 0.533 g/cm^3. And the densest metal? Osmium (Os) at a whopping 22.587 g/cm^3.

Pure silver doesn't occur in nature. It has to be refined to separate it from other elements it's found with.

CHAPTER THREE

WHERE DOES SILVER COME FROM?

Silver has been mined all over the earth for thousands of years. In ancient times, silver often came from the vicinity of modern-day Turkey. Later, deposits were found in Europe in what's now Poland and Germany. In the 16th century, Spanish explorers found rich deposits in South America, and miners in the days of the California Gold Rush also found veins of silver.

Silver is the 66th most abundant element on Earth. It tends to be found near deposits of other metals such as gold or copper. Silver ore can also be found near zinc and lead deposits. In fact, about half the silver that's mined is found while mining for other metals.

The process of removing silver ore from the rock it's found in is called smelting. This involves heating the ore beyond the melting point of the metal being extracted. For silver, this requires heat greater than 1,763.2°F (961.8°C).

MORE PRECIOUS THAN GOLD?

The first known silver mines were in Asia Minor (modern-day Turkey) and Greece nearly 5,000 years ago. Silver from these mines was worked by skilled craftspeople into jewelry, bowls, cups, and other objects. Because silver was desired by many ancient cultures, the groups or nations that controlled the sources of silver often became very wealthy and powerful.

Ancient Egyptians valued silver because it had to be imported from distant lands, unlike gold, which could be found nearby.

Today, silver coins are produced only as collectors' items, but in the ancient world they were used as everyday currency.

About 600 BCE, large silver deposits were discovered near Athens, Greece. These became known as the Laurium mines. Their silver helped build the Greek empire. Like the Greeks, the Roman people also used silver to finance their empire. Most of their silver came from mines in what is now Spain, over which they had gained control after winning the Second Punic War in 202 BCE. With ready sources of silver, both the Greeks and the Romans used silver coins. The ancient Greeks used the drachma, while the Romans minted a coin called the denarius. These coins ranged in weight from one-eighth to one-seventh of an ounce (or three and a half to four grams).

The Spanish mines controlled by the Romans were an important source of silver for nearly 1,000 years. Additional large deposits were discovered in Germany and eastern Europe during medieval times (800 to 1200 CE). The silver from these mines was traded all over Europe, Asia, and North Africa. In China, the silver was often traded for the unique spices and silk that many Europeans craved.

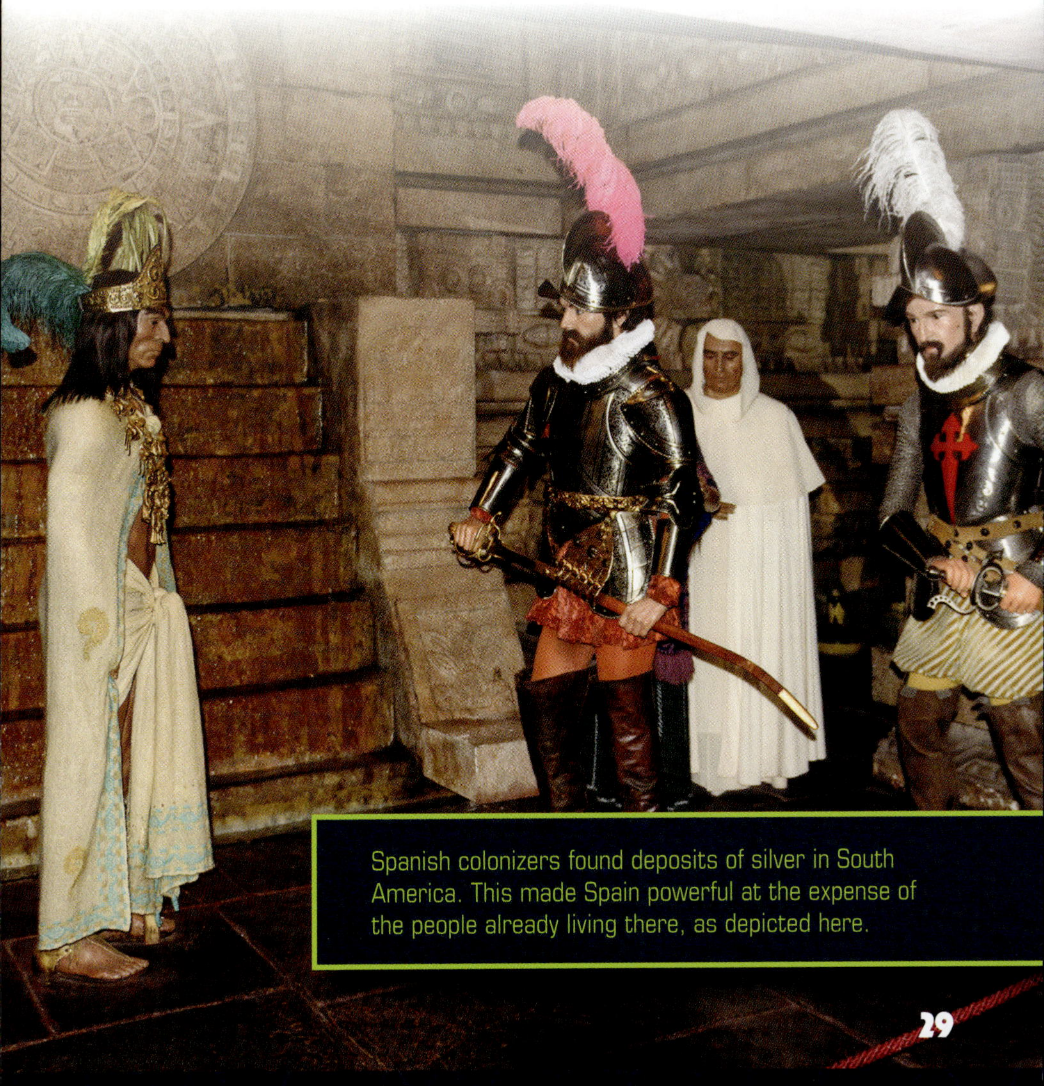

Spanish colonizers found deposits of silver in South America. This made Spain powerful at the expense of the people already living there, as depicted here.

SILVER FROM THE AMERICAS

In 1492, Christopher Columbus (1451–1506) landed in the Americas and claimed land for the king of Spain. He was followed soon after by Spanish soldiers and explorers known as conquistadores. These men were driven by the desire to find silver and gold. One conquistador was Francisco Pizarro (about 1475–1541). His army conquered the Inca, the Native people of Peru, and claimed their stores of gold and silver. Pizarro's victory, however, would cost him his life. In 1541, he was murdered by Spanish soldiers who wanted a larger share of his treasure.

The conquistadores discovered large silver deposits throughout Central and South America. One of the most profitable was the Potosi mine in today's Bolivia, which shipped millions of ounces of silver to the king of Spain. For the next 300 years, mines in Bolivia, Peru, and Mexico produced most of the world's silver.

In the mid-19th century, the U.S. state of Nevada would become a hot spot for silver. Rich deposits like the Comstock lode would draw hundreds of miners and prospecting companies seeking to strike it rich. Smaller mines were also discovered in Utah and Colorado. About this time, advances in technology made it less dangerous and easier to mine silver.

WHERE DOES SILVER COME FROM NOW?

According to the Silver Institute, an international association of people in the silver industry, more than 830 million ounces (23.5 million kg) of silver were mined in 2023. That was a decrease of about 6 million ounces (270,000 kg) from the previous year. As it has been since the 16th century, much of the world's silver comes from Central and South America.

The biggest producer of silver is Mexico, followed by China. Silver has always been highly valued in Chinese culture. Currently there are 927 silver mining operations in China, making it the number two producer of silver in the world. China accounts for 13 percent of global silver production.

Because silver has to be separated from other minerals, silver refining is often part of other mineral mining operations.

TOP TEN SILVER-PRODUCING COUNTRIES IN 2023

	MILLIONS OF OUNCES	MILLIONS OF KILOGRAMS
MEXICO	202.2	5.73
CHINA	109.3	3.1
PERU	107.1	3.04
CHILE	52	1.47
BOLIVIA	42.6	1.21
POLAND	42.5	1.20
RUSSIA	39.8	1.13
AUSTRALIA	34.4	2.04
UNITED STATES	32	0.91
ARGENTINA	26	0.74

A smelting furnace reaches temperatures of thousands of degrees. Smelting is the process of melting metal ore out of the rocks it formed in.

The Parkes process uses zinc to separate precious metals from lead.

GETTING SILVER FROM LEAD

Alchemists in medieval times were obsessed with finding a way to turn lead into gold. While that's the stuff of fantasy novels today, they might be surprised to learn that much of today's silver comes from lead after going through what is known as the Parkes process. This process involves the use of zinc to separate the silver from lead. When zinc and lead are melted and stirred together, the zinc forms compounds with the silver and can then be separated.

Silver is found in four forms in nature: as a native element, as part of an alloy, as part of silver minerals, or as a byproduct of another metal ore.

Native silver doesn't form the type of crystals common in other minerals. Instead, it forms flakes, plates, and sometimes natural "wire." Silver naturally forms alloys with other metals, including gold. An alloy of gold and silver with at least 20 percent silver is called electrum. Silver is also found as part of 39 different minerals, most of which are too rare to be mined economically.

Most silver used today is found as part of other ores. Silver is often obtained as a byproduct of mining copper, zinc, or lead. Sometimes the silver is even more valuable than the primary metal being mined!

The Parkes process was patented by Alexander Parkes in 1850. You could call him a modern-day alchemist!

CHAPTER FOUR

SILVER MAKES FRIENDS

A substance made of only one kind of atom is an element. A substance made of more than one kind of atom bonded together is called a compound. Elements can form two different types of bonds—covalent or ionic. A covalent bond is formed when atoms share pairs of electrons.

An ionic bond is formed when electrons from the valence shell of one atom transfer to another atom. This forms a molecule of an entirely different substance. For example, silver has one electron in its outer shell, while iodine has seven. Atoms with a filled outer shell of electrons are more stable. When an iodine atom comes near a silver atom, it pulls away silver's lone outer electron. By gaining one electron, an iodine atom gets a filled outer shell and becomes more stable. This new substance is called silver iodide.

When silver loses an electron, it forms a cation, or a positively charged ion. When iodine gains an electron, it becomes an anion, or a negatively charged ion. The two oppositely charged ions are attracted to one another. This is what forms an ionic bond.

SOMETHING + SOMETHING = SOMETHING ELSE

Chemical formulas are a shorthand way of describing compounds. A chemical formula indicates which atoms are bound together and in what ratio. For example, the formula for silver iodide is AgI. Ag is the chemical symbol for silver, while I is the chemical symbol for iodine. The elements combine in an atomic ratio of one to one. In other words, one atom of silver combines with one atom of iodine.

Ag_2S is the chemical formula for a compound called silver sulfide. The "2" to the right of the symbol for silver indicates that there are two atoms of silver (and one atom of sulfur) in each molecule of silver sulfide. Silver sulfide will be discussed in more detail later in this chapter.

SILVER IODIDE

Silver iodide is used for photography, as an antiseptic, and for cloud seeding.

Chemical reactions are described by chemical equations. Let's look at the reaction between silver and iodine to form silver iodide. Iodine is represented as I_2 because it has two atoms in a molecule, and silver is represented as Ag because it's not molecular. The chemical equation for this reaction is:

$$2\,Ag\,(s) + I_2\,(g) \rightarrow 2\,AgI\,(S)$$

In all chemical equations, the substance or substances to the left of the arrow are known as the reactants, while the substance or substances to the right of the arrow are called the product. The equation shows that two atoms of silver combine with two atoms of iodine to form silver iodide. The "s" indicates that the silver and silver iodide are solids. The "g" indicates that the iodine is a gas (at room temperature, iodine is a solid).

SILVER + HALOGEN

Silver halides are ionic compounds made of silver and a halogen. (A halogen is any of the elements from group VIIA of the periodic table.) The three most important silver halides are silver bromide (AgBr), silver chloride (AgCl), and the previously described silver iodide. All ionic compounds of silver are, to some degree, sensitive when exposed to light. All three of these silver compounds are used to make photographic and X-ray film. When light strikes this film, it causes the silver halide to break down and produce a speck of metallic silver.

SILVER ALLOYS

Mixtures of two or more metals are called alloys. Because silver is a relatively soft metal, most things made of silver are alloys—mostly silver with small amounts of harder, usually less expensive, metals. For instance, pure silver flatware would bend too easily to be useful. Silver flatware was generally made of sterling silver. The copper content makes the silver harder.

FINE SILVER	99.9 PERCENT PURE SILVER	THE PUREST SILVER YOU CAN BUY
STERLING SILVER	92.5 PERCENT SILVER AND 7.5 PERCENT COPPER	COMMONLY USED FOR JEWELRY AND ORNAMENTAL OBJECTS
ARGENTIUM, SILVADIUM, STERLIUM, STERILITE	92.5 TO 96 PERCENT SILVER, WITH THE REST COPPER, ZINC, AND OTHER METALS	THESE ALLOYS ARE MARKETED AS TARNISH RESISTANT
BRITANNIA SILVER	95.83 PERCENT SILVER, THE REMAINDER COPPER	A LEGAL STANDARD FOR SILVER PRODUCED IN ENGLAND IN THE 17TH CENTURY
COIN SILVER	90 PERCENT SILVER AND 10 PERCENT COPPER	AT ONE TIME, THE STANDARD FOR U.S. SILVER COINS

X-ray images are captured on film coated with silver compounds. There are companies that specialize in retrieving the silver from old X-ray film so it can be recycled.

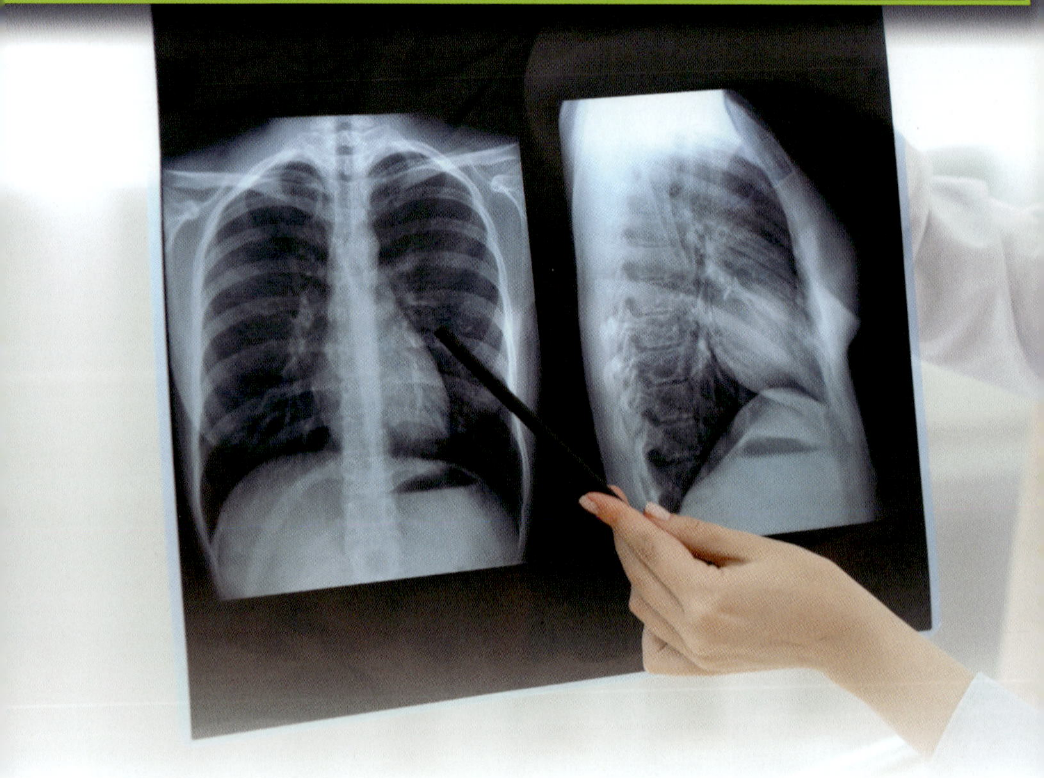

The film is then developed to bring out the image. The process of developing a silver exposure involves a chemical reaction in which all of the silver halide is converted to silver metal. In essence, a photograph is an image drawn with silver.

Silver iodide is also used to produce rain artificially. This procedure, which is known as cloud seeding, has been used in areas that are stricken by drought and in desperate need of rain. It can help produce precipitation that gives people the water they need to survive.

In cloud seeding, people in airplanes spray silver iodide into a cloud. The silver iodide causes

Cloud seeding is most often used to increase rain or snowfall. However, it was also be used to alter precipitation. Silver iodide can be introduced to hail clouds to reduce the size of ice particles. It can also be used to disperse fog and improve visibility.

water droplets to form ice crystals, which fall out of the cloud. As the ice crystals fall, they melt and turn into rain

SILVER + NITRIC ACID

Silver nitrate ($AgNo_3$) is a colorless compound that is made by reacting silver with nitric acid (HNo_3). It is poisonous and must be handled very carefully. Silver nitrate is commonly used in the preparation of silver halides for photographic film. The equation below shows how silver nitrate reacts with table salt (NaCl) to create silver chloride. The "aq" in parentheses stands for "aqueous," which means that the compounds are dissolved in water.

$$AgNO_3\,(aq) + NaCl\,(aq) \rightarrow AgCl\,(s) + NaNO_3\,(aq)$$

The reaction results in solid silver chloride gathering on the bottom of the container in which the reactants are mixed. The formation of a solid when two liquids are mixed is known as a precipitation reaction.

In the past, eye drops of 1 percent silver nitrate solution were used to prevent eye infections in newborns. This is because silver nitrate kills bacteria. Today, antibiotics, which are inexpensive and readily available, are more commonly used for this purpose.

SILVER + SULFUR

Silver tarnishes, or turns brownish black, when sulfur compounds in the atmosphere react with the silver surface to form black silver sulfide. It takes special polish and a lot of elbow grease to return silver to its shiny glory. This is one reason why silver tableware isn't as popular today.

When an object made of silver is exposed to air long enough, it loses its shine and turns dull gray or black. This is called tarnishing. Silver objects such as a fork or a knife tarnish because they react with hydrogen sulfide (H_2S) in the air to create a thin layer of silver sulfide (Ag_2S). Some foods, including mustard and eggs, contain sulfur and can tarnish silver objects. If you use a silver knife to put mustard on your sandwich, be sure to wash it off quickly. If you don't, the knife might become coated with an unappealing layer of silver sulfide.

Argentite is silver sulfide that occurs naturally in Earth's crust. It is also known as acanthite. The Comstock lode was a famous deposit of argentite in Nevada that was mined for its silver content.

SILVER, SILVER EVERYWHERE

Silver is in many forms all around us, every day. Silver ions are used in water purification systems, both on Earth and in outer space. They have been used for water purification on the Apollo spacecraft and the International Space Station. Silver is also used in ice machines to keep mold, bacteria, and algae from forming.

Silver coatings are used for medical devices like catheters because of their antimicrobial properties. During World War I, medics used silver for battlefield dressings and to suture wounds. Today, researchers are investigating the use of silver to combat antibiotic-resistant “superbugs.”

CHAPTER FIVE

SILVER IN OUR LIVES

There is silver everywhere in our world. Some of it is obvious, like silver jewelry or fancy candlesticks. Other silver might not be as visible, but it's equally important. Silver conducts electricity very well, so it's used to make wires and circuits. There's silver inside your computer and cell phone. When you watch old movies, you're seeing the silver halides used in film at work. Let's learn about how silver impacts our day-to-day lives.

SILVER IN YOUR KEYBOARD

Of all the elements, silver is the best conductor of electricity. It is such a good conductor that very little silver is needed to pass an electrical current. Used in electronic equipment, a tiny bit of silver can go a long way. For example, beneath every key in a computer keyboard are two thin patches of silver separated by a tiny space, which is called a silver membrane switch. When a key is pressed, the two patches touch, sending an electrical current that races toward the computer's central processing unit. The result is a letter appearing on the computer screen. Silver is ideal for this task because it is extremely sensitive. It only takes a light touch of the key to produce the current. Silver is also durable. You can hit the key again and again, and the silver contact will not fail. Silver membrane switches are used in many other machines. According to the Silver Institute, "Every time a homeowner turns on a microwave oven, dishwasher, clothes washer, or television set, the action activates a switch with silver contacts that completes the required electrical circuit."

Your grandparents or other older relatives may have received a tea set like this as a wedding present. Today, the silver in your home is much more likely to be hidden deep inside your electronic devices—or even the wall switch for your lights!

Can you imagine your day-to-day life without a microwave oven or cell phone? Silver in the circuit boards makes those things function.

If you have ever opened a computer, you have probably seen a circuit board. It is a thin green piece of plastic with small circular and rectangular pieces stuck to it. On the surface of the circuit board is a mazelike pattern of lines. These devices are also found in mobile phones, electrical appliances, and televisions. Circuit boards often use silver. A silver alloy known as solder is used as a glue to attach the components to the board and ensure that electricity flows freely. Because silver is such a great conductor of electricity, it takes just a tiny bit of it to get the job done.

FINENESS VS. PURITY

Fineness and purity are ways of describing how much silver an object contains. They're similar but not the same. Purity is the percentage of pure silver an object contains. Fineness is the amount of silver per thousand. For example, if a silver alloy has a fineness of 999, this means, in terms of purity, it is 99.9 percent pure silver, with only 0.1 percent other metals.

Fineness and purity describe the same thing in different ways.

This American Eagle 1-ounce (28.4 g) silver dollar coin sells for $95—considerably more than the face value! It's marketed to collectors and not used as legal tender.

SILVER COINS

Silver was first used for coins more than 2,500 years ago. Silver's properties make it an ideal metal for coins. Because it is malleable, it can be shaped into a thin, round coin that is easily carried. It is also soft enough that a design can be stamped on its surface. This design might identify the nation that issued the coin, and it also designates the coin's value. For much of human history, silver was the metal of choice for coins.

Today, however, all nations have removed their silver coins from circulation. This is because silver has become too expensive. The silver content in a coin, if made today, would be worth more than the face value of the coin. In the United States, silver coins were last issued in 1965. In England, silver coins were last minted by the government in 1947. Although silver coins are not used to make purchases anymore, many countries release commemorative silver coins to honor famous people, places, or events. These coins are popular with collectors, and most of them will increase in value over the years.

WHAT CAN YOU MAKE OUT OF SILVER?

Silver is an ideal metal for jewelry. It is soft enough to be shaped and stretched, yet durable enough to withstand regular wear and tear. It can also be polished to create a reflective surface. Mixed with other metals to create alloys, silver's hardness and color can be customized to meet the needs of the artist.

People have made jewelry out of silver for centuries. It's still popular today.

WHAT'S SILVER WORTH?

Like gold or stocks, silver is bought and sold in an international market. Some companies buy silver to use in manufacturing. Others buy it as an investment, intending to sell later at a higher price. Silver is priced by ounce, gram, and kilogram. In late January 2025, silver was selling at about $31 per ounce.

Jewelry stores and antique shops will often buy sterling silver jewelry or decorative objects. Depending on the type of shop, the owner may pay market price for the weight of the silver, with the intent of melting it down. Or, they may offer a different price based on the current value of the item in the antique market.

Today's collectible silver coins are prized for their shining, detailed images. They're usually sold sealed with a cover to prevent tarnish.

Fancy silver tableware used to be passed down in families from generation to generation as heirlooms. However, most people today don't have time to spend polishing knives and forks.

Forks, knives, and spoons made of sterling silver are highly valued for their beauty. However, since silver isn't the most affordable metal, a lot of silverware today is made of less-expensive metal. Silver-plated utensils are also popular. Silver plating is the process in which an object made of less expensive metal is coated, or plated, with a thin layer of silver. This is done by placing the object in a solution of silver cyanide, along with a bar of silver. The object to be plated and the bar of silver are connected to a battery. This causes the silver bar to dissolve, forming silver ions to replace those that were deposited on the object, and the silver ions to flow through the solution and deposit themselves on the object.

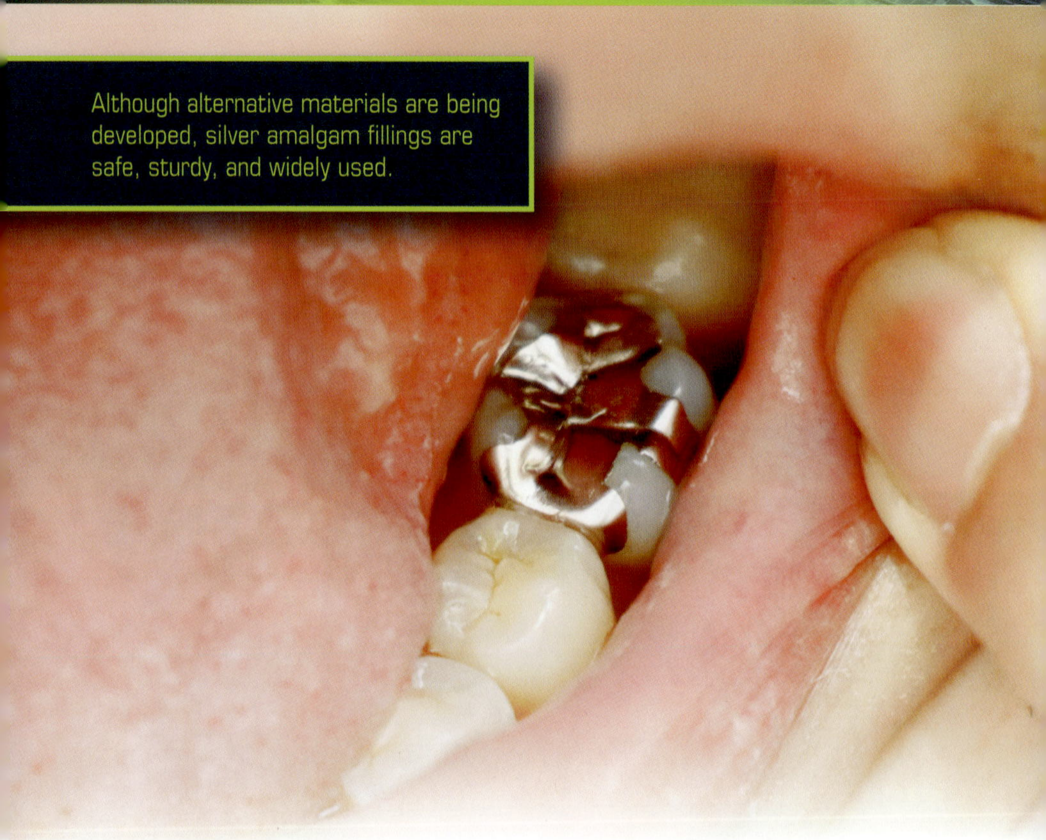

Although alternative materials are being developed, silver amalgam fillings are safe, sturdy, and widely used.

SILVER IN YOUR TEETH

Silver is used to fill cavities in teeth. The filling isn't pure silver, which would be too soft, but a mixture of silver, mercury, tin, copper, and sometimes zinc. This mixture is known as an amalgam. The amalgam is a soft paste that, when applied to a tooth, quickly hardens and expands to create a durable filling.

Silver amalgam fillings have become less popular in recent years because of concerns that the mercury in the filling is harmful to the body. However, as of the past few years, the U.S. Food and Drug Administration and the American Dental Association have stated that they believe amalgam fillings are perfectly safe.

A SILVER LINING

People have valued silver for thousands of years for many different reasons. Ancient people prized silver for its shine and beauty, and its ability to be worked into different shapes, such as in jewelry. For many centuries, it was the material of choice for coins. Later, it was a symbol of wealth and prestige to have silver trays and tableware. A silver watch or decorative object was long considered a fitting gift for a major milestone like a wedding or retirement—and, of course, the silver, or 25th, wedding anniversary.

In today's fast-paced society, people don't value decorative objects that need to be polished to look nice as much as previous generations did. However, silver still is all around us. It's inside the electronic devices we depend on, the filtration systems that keep our water safe, and often in many jewelry boxes.

Today's wristwatches may not look like your great-grandpa's pocket watch, but a silver watch is still considered a classic gift for important milestones such as graduation.

THE PERIODIC TABLE OF ELEMENTS

Unlike many other chemical elements, silver is something tangible that most people see and handle in some form in their daily lives.

PERIODIC TABLE OF ELEMENTS

- Alkali metals
- Alkaline earth metals
- Transition metals
- Metal
- Metalloids
- Nonmetals
- Halogen
- Noble gases
- Lanthanoids
- Actinoids

1 H Hydrogen 1.008																	2 He Helium 4.0026
3 Li Lithium 6.94	4 Be Beryllium 9.0122											5 B Boron 10.81	6 C Carbon 12.011	7 N Nitrogen 14.007	8 O Oxygen 15.999	9 F Fluorine 18.998	10 Ne Neon 20.180
11 Na Sodium 22.990	12 Mg Magnesium 24.305											13 Al Aluminium 26.982	14 Si Silicon 28.085	15 P Phosphorus 30.974	16 S Sulfur 32.06	17 Cl Chlorine 35.45	18 Ar Argon 39.948
19 K Potassium 39.098	20 Ca Calcium 40.078	21 Sc Scandium 44.956	22 Ti Titanium 47.867	23 V Vanadium 50.942	24 Cr Chromium 51.996	25 Mn Manganese 54.938	26 Fe Iron 55.845	27 Co Cobalt 58.933	28 Ni Nickel 58.693	29 Cu Copper 63.546	30 Zn Zinc 65.38	31 Ga Gallium 69.723	32 Ge Germanium 72.630	33 As Arsenic 74.922	34 Se Selenium 78.971	35 Br Bromine 79.904	36 Kr Krypton 83.798
37 Rb Rubidium 85.468	38 Sr Strontium 87.62	39 Y Yttrium 88.906	40 Zr Zirconium 91.224	41 Nb Niobium 92.906	42 Mo Molybdenum 95.95	43 Tc Technetium (98)	44 Ru Ruthenium 101.07	45 Rh Rhodium 102.91	46 Pd Palladium 106.42	47 Ag Silver 107.87	48 Cd Cadmium 112.41	49 In Indium 114.82	50 Sn Tin 118.71	51 Sb Antimony 121.76	52 Te Tellurium 127.60	53 I Iodine 126.90	54 Xe Xenon 131.29
55 Cs Caesium 132.91	56 Ba Barium 137.33	57-71	72 Hf Hafnium 178.49	73 Ta Tantalum 180.95	74 W Tungsten 183.84	75 Re Rhenium 186.21	76 Os Osmium 190.23	77 Ir Iridium 192.22	78 Pt Platinum 195.08	79 Au Gold 196.97	80 Hg Mercury 200.59	81 Tl Thallium 204.38	82 Pb Lead 207.2	83 Bi Bismuth 208.98	84 Po Polonium (209)	85 At Astatine (210)	86 Rn Radon (222)
87 Fr Francium (223)	88 Ra Radium (226)	89-103	104 Rf Rutherfordium (267)	105 Db Dubnium (268)	106 Sg Seaborgium (269)	107 Bh Bohrium (270)	108 Hs Hassium (277)	109 Mt Meitnerium (278)	110 Ds Darmstadtium (281)	111 Rg Roentgenium (282)	112 Cn Copernicium (285)	113 Nh Nihonium (286)	114 Fl Flerovium (289)	115 Mc Moscovium (290)	116 Lv Livermorium (293)	117 Ts Tennessine (294)	118 Og Oganesson (294)

57 La Lanthanum 138.91	58 Ce Cerium 140.12	59 Pr Praseodymium 140.91	60 Nd Neodymium 144.24	61 Pm Promethium (145)	62 Sm Samarium 150.36	63 Eu Europium 151.96	64 Gd Gadolinium 157.25	65 Tb Terbium 158.93	66 Dy Dysprosium 162.50	67 Ho Holmium 164.93	68 Er Erbium 167.26	69 Tm Thulium 168.93	70 Yb Ytterbium 173.05	71 Lu Lutetium 174.97
89 Ac Actinium (227)	90 Th Thorium 232.04	91 Pa Protactinium 231.04	92 U Uranium 238.03	93 Np Neptunium (237)	94 Pu Plutonium (244)	95 Am Americium (243)	96 Cm Curium (247)	97 Bk Berkelium (247)	98 Cf Californium (251)	99 Es Einsteinium (252)	100 Fm Fermium (257)	101 Md Mendelevium (258)	102 No Nobelium (259)	103 Lr Lawrencium (266)

- C Solid
- Hg Liquid
- H Gas
- Rf Unknown

GLOSSARY

alloy: A mixture of two or more metals.
argentite: A valuable silver ore, also known as silver sulfide.
atom: The smallest unit of an element. The atom itself is made of electrons, protons, and neutrons.
compound: A substance made up of two or more elements bound together by chemical bonds. The elements in a compound combine in fixed ratios, such as two atoms to one, or three to two, etc.
conduct: To allow something, such as heat or electricity, to pass through.
current: The flow of electrically charged particles.
density: The mass of a sample divided by its volume.
ductile: Capable of being stretched into a wire.
electron: A negatively charged particle found outside of the nucleus, or center, of an atom.
fineness: A measure of the purity of silver expressed as parts per thousand.
ion: A positively or negatively charged atom or group of atoms.
lode: An area that contains large amounts of metal sandwiched between layers of rock.
malleable: Capable of being bent or shaped.
matter: Anything that has mass and exists as a solid, liquid, or gas.
molecule: Two or more atoms joined together by chemical bonds. A molecule is the smallest particle of a substance that still retains its chemical properties.
neutron: A particle without charge that is part of the nucleus of most atoms.

ore: A rock that contains a valuable metal.
precipitation: The release of a substance that has been dissolved in a solution in solid form.
proton: A positively charged particle that is part of the nucleus of an atom.
radioactive: Capable of releasing high-energy rays or particles.
room temperature: The average indoor temperature at which experiments are performed. Scientists consider room temperature to be approximately 68°F (20°C).
talc: A very soft mineral also known as magnesium silicate.
tarnish: Discoloration of metal due to a chemical reaction.
valence electron: An electron in the outer shell of an atom.

FOR MORE INFORMATION

AMERICAN MUSEUM OF NATURAL HISTORY

200 Central Park West
New York, NY 10024-5102
(212) 769-5000
Website: www.amnh.org
Facebook: @naturalhistory
Instagram and X: @amnh
AMNH is a museum dedicated to advancing research and sharing scientific knowledge about the natural world.

CHEMICAL INSTITUTE OF CANADA (CIC)

90-2420 Bank Street
Ottawa, Ontario, Canada
K1V 8S1
(613) 232-6252
Website: www.cheminst.ca/
X: @CIC_ChemInst
The CIC facilitates the research and cooperation of chemists and chemical engineers in Canada.

INTERNATIONAL UNION OF PURE AND APPLIED CHEMISTRY (IUPAC)

IUPAC Secretariat
79 T.W. Alexander Drive
Research Commons Building 4501, Suite 190
Research Triangle Park, NC 27709, USA
(919) 485 8700
Website: iupac.org/
Facebook: @iupac.org
X: @iupac
IUPAC is the global chemistry organization that maintains the periodic table of elements.

INTERNATIONAL UNION OF PURE AND APPLIED PHYSICS (IUPAP)

Fondazione Internazionale Trieste per il Progresso e la Libertà delle Scienze
c/o ICTP
Strada Costiera 11
34151 Trieste
Website: iupap.org/
Facebook: @iupap
X: @iupap_physics
IUPAP's mission is to assist in the global advancement of physics and facilitate international scientific cooperation.

THE SILVER INSTITUTE

1400 I Street NW, Suite #550
Washington, DC 20005
(202) 835-0185
Website: silverinstitute.org
The Silver Institute is a global resource for the uses and market value of silver.

FOR FURTHER READING

Congdon, Lisa. *The Illustrated Encyclopedia of the Elements: The Powers, Uses, and Histories of Every Atom in the Universe*. San Fransico, CA: Chronicle Books, LLC, 2021.

DK. *Super Simple Chemistry: The Ultimate Bitesize Study Guide*. New York, NY: DK Publishing, 2020.

Huggins-Cooper, Lynn. *Chemistry for Curious Kids: An Illustrated Introduction to Atoms, Elements, Chemical Reactions, and More!* London, UK: Arcturus, 2021.

McHenry, Ellen Johnston. *The Chemical Elements Coloring & Activity Book: A Fun and Interactive Guide to the World of Chemistry*. Independently Published, 2021.

Silver, Donald, and Patricia Wynne. *My First Book About Chemistry*. Mineola, NY: Dover Publications, 2020.

Thomas, Isabel, and Sara Gillingham. *Exploring the Elements: A Complete Guide to the Periodic Table*. New York, NY: Phaidon Press, Inc., 2021.

Woodbury, Rebecca. *Atoms and Molecules Meet*. Real Science-4-Kids, 2024.

Zovinka, Edward P. and Rose A. Clark. *A Kids' Guide to the Periodic Table: Everything You Need to Know About the Elements*. Emeryville, CA: Rockridge Press, 2020.

INDEX

INDEX

ABOUT THE AUTHOR

KATHLEEN A. KLATTE is the author of many nonfiction books for children and teens. Topics she has written about range from animals and nature to constitutional law to unusual career choices. She lives in New York with one cat and far too many books and Legos.

PHOTO CREDITS

Cover, p. 1 remotevfx.com/Shutterstock.com; Cover, pp. 1, 3-64 pluie_r/Shutterstock.com; Cover, pp. 1, 3-64 Oksancia/Shutterstock.com; p. 5 Everett Collection/Shutterstock.com;p . 6 Nor Gal/Shutterstock.com; p. 7 VladKK/Shutterstock.com; p. 8 JLStock/Shutterstock.com; p. 9 vchal/Shutterstock.com; p. 11 Phawat/Shutterstock.com; p. 13 isak55/Shutterstock.com; p. 15 Srikanth_K/Shutterstock.com; p. 17 Vera Kalyuzhnaya/Shutterstock.com; p. 19 Irina Mosina/Shutterstock.com; p. 20 R-studio/Shutterstock.com; p. 21 Yasni/Shutterstock.com; p. 23 Nordroden/Shutterstock.com; p. 25 RHJPhtotos/Shutterstock.com; p. 27 CK-TravelPhotos/Shutterstock.com; p. 28 Alex Coan/Shutterstock.com; p. 29 Anton_Ivanov/Shutterstock.com; p. 31 Koshiro K/Shutterstock.com; p. 33 Mark Agnor/Shutterstock.com; p. 34 Bjoern Wylezich/Shutterstock.com; p. 35 RHJPhtotos/Shutterstock.com; p. 37 Peter Hermes Furian/Shutterstock.com; p. 40 New Africa/Shutterstock.com; p. 41 zmpixes/Shutterstock.com; p. 42 Foolish Moments/Shutterstock.com; p. 45 David Huntley Creative/Shutterstock.com; p. 46 aPhoenix photographer/Shutterstock.com; p. 47 VladKK/Shutterstock.com; p. 48 Ayman Noureldin/Shutterstock.com; p. 49 Pixel-Shot/Shutterstock.com; p. 50 Adam F Barner/Shutterstock.com; p. 51 hanohiki/Shutterstock.com; p. 52 umaruchan4678/Shutterstock.com; p. 53 Chris Haye/Shutterstock.com; p. 54 ShannonChocolate/Shutterstock.com; p. 55 View_From_My_Lens/Shutterstock.com.